THE INCREDIBLE HUMAN BODY

Ilene L. Follman

Advisors

Karen Hales Mecham

Ann C. Edmonds

Primary Science Resource Guide
Includes Transparencies, Reproducibles, and Teacher's Guide

THE INCREDIBLE HUMAN BODY

Illustrations: Maggie Rechtiene, Nadine Sokol
Project Manager: E. Rohne Rudder
Cover Design: E. Rohne Rudder
Typesetting: Pagecrafters, Inc.
Managing Editor: Kathleen Hilmes

To the Teacher

A marvelous machine . . . a chemical laboratory, a power-house.
Every movement, voluntary or involuntary, full of secrets and marvels!

Theodor Herzl
Dictionary of Quotable Definitions

As young children grow, their bodies change in both dramatic and subtle ways. Obvious signs of growth and change are reflected in outgrown clothes that must be replaced and in the loss of baby teeth to make room for permanent teeth. More subtle are the internal changes occurring as bones lengthen, muscles increase in mass and strength, and brain cells process increasingly more complex information.

The information and activities in this resource book enhance the child's knowledge and awareness of the many parts of the body responsible for human growth, health, and life-sustaining activities. The book is organized roughly into three parts: the **outside** of the body, the **inside** of the body, and **caring** for the body. Children will learn about the functions of major organs and systems of the body. They will explore ways in which the sense organs help them learn about their surroundings. They will become aware of how the body protects itself and how it expresses emotion through gestures, tears, and laughter. And they will learn about how they can maintain and protect their body through proper nutrition, exercise, rest, hygiene, and safety.

The material follows a **learning cycle** that begins with: **a) free exploration** and discovery by students; **b) guided expansion of explorations** through activities that allow students to test, integrate, and sort out their discoveries; and **c) application of concepts**, accomplished through individual and group projects that give students the opportunity to consolidate learning and to share their learning with others. Emphasis includes the curricular areas of science, math, art, social studies, reading, writing, drama, and movement. Activities involve skills in observing, classifying, measuring, recording, predicting, listening, writing, brainstorming, constructing, explaining, describing, comparing, contrasting, organizing, and sharing.

A *To The Parent* reproducible page gives a brief explanation of the unit and suggests ways parents can share in or enhance some of the children's activities and discoveries.

Four **reproducible pages** are included to actively involve students in recording individual observations and discoveries. They are referenced as **R** within the text.

Four **transparencies** engage students in discussion and understanding about how the systems and parts of the body fit together into a whole. They are referenced as **T** within the text. Spaces are provided so the class can write in the names of the organs indicated. Transparency instructions are printed on the inside front and back covers.

A **bibliography** includes books and resources for children and teachers.

Contents

To the Parent

During the next several weeks, your child will be participating in activities that focus on the human body. Students will learn about the **outside** of the body, including the functions of skin and hair, and the importance of sight, hearing, smell, taste, and touch. They will study the **inside** of the body, focusing on bones, muscles, and major organs of digestion, respiration, and circulation. They will become more aware of how the body protects itself, and how it expresses emotions through gestures, tears, and laughter. They will learn about the body's need for proper nutrition, exercise, rest, hygiene, and safety.

Your child will discover many new things about human growth, change, and life-sustaining activities that occur within the body every moment of the day. You can be part of that discovery process by:

• talking with your child about some of the discoveries made in class;

• reading books or stories with your child about various facets of the human body;

• talking with your child about the qualities you both share as human beings and the ways in which each of you is a unique individual;

• helping your child prepare a nutritionally balanced breakfast, lunch, or snack;

• asking your child to help you put together a small, household first-aid kit; discussing reasons for household health routines and safety rules;

• sharing feelings or anxieties about procedures in a doctor's office or hospital if your child is ill or injured; discussing the reasons for visiting a doctor or dentist.

Getting Started

Students can find out much about their own bodies by talking about daily habits and routines, sharing experiences about visits to the doctor, dentist, or hospital, looking at themselves or others in photographs or in a full length mirror, or engaging in strenuous physical activities.

1) Talk informally with the class about a typical day in their lives, including bedtime and waking routines, meals, and outdoor activities. What parts of their daily routines affect—or are affected by—the children's bodies? They will soon discover that the answer is "All!"

2) If a child is ill or has been injured in some way, discuss what might help the child feel better again: care by a school nurse, a doctor, or hospital staff; bandages; medicine; rest or sleep. Assess students' fears, knowledge, and depth of experiences concerning their own bodies. *Use their responses and your assessment in developing the scope and pace of further activities.*

3) Establish a display and discovery area for free exploration. Equip the area with a magnifying glass, microscope, mirror (full length, if possible), photographs of the students, books and stories about the human body, a stethoscope, bones (from a chicken, turkey, or other animal), and a first-aid kit. Change or add to the area to introduce new concepts and share materials created or used by students in activities centered around the human body.

4) Take photographs of each child, or ask the parents to provide a recent photo. Mount the photos on a large piece of tagboard, and place it in the display area. Talk about characteristics the students observe about one another. What characteristics allow students to recognize one another? Clothing? Hair color or style? Eye color? Sounds of voices?

5) Transparencies A, B, C, and **D** can be used separately to introduce, discuss, or summarize an organ or system of the body. They can also be used **together** to give a more complete picture of the inside of the body. See the inside covers for additional transparency instructions.

6) Transparency A, the skeleton, is divided into separate parts: head, trunk/torso, upper legs, lower legs, upper arms, lower arms, feet, and hands. The skeleton can also be laid on top of **Transparencies B, C,** and **D** to show children how bones fit with other systems.

Keeping Records

1) Use the **reproducible pages (R)** included in this book as recording devices.

2) Provide individual **portfolios** (e.g., file folders) for students to save drawings, reproducible pages, written work, or other materials pertinent to their study of the human body.

3) Provide **journals** in which students can make daily entries about observations, discoveries, thoughts, and other information about their own bodies. Journals can consist of several loose-leaf, lined pages held together by colored yarn looped through margin holes and tied. **Include journal entries as an integral part of an activity**; do not wait until the end of the school day. Younger students may dictate journal entries to the teacher or draw pictures as their entries. Initial journal entries may be very simple, amounting to no more than a descriptive word, phrase, or drawing. As the school year progresses, recording will become more sophisticated.

4) Create a **class book** about the human body. Ask students to select some of their individual portfolio materials to copy and include in the book. Include charts, photos of students, drawings, new discoveries or conclusions, and other information the class feels is important (see page 31, *Making Connections*).

The Incredible Human Body

The Outside of the Body: *Things About Me*

Identifying and discovering more about *visible* body characteristics is an easy way for students to begin studying the human body. The activities below emphasize that although students may share the same general characteristics, each child is unique according to the way in which those characteristics manifest themselves. For example, eyes have irises of different colors. Feet, hands, arms, and legs vary in size or length. Several students may have brown hair, but each may differ by being thick or thin, curly or straight, or long or short.

What is hair? In what ways is all human hair the same? Different? Hair grows out of the skin, covering much of the body. Each hair consists of a long **shaft** of dead material, and a **roo**t of live cells enclosed in a light-colored bulb nestled in a **follicle** below the skin. Live cells in the root add new material to the hair shaft, causing the hair to grow. The shape of the follicle determines whether the hair on the head will be straight (round follicle), wavy (oval follicle), or curly (flat follicle). A muscle attached to the follicle raises the hair—makes it "stand on end"—when a person is cold or frightened. Hair color comes from a pigment called **melanin**, which is deposited in hair cells as they form at the root of the hair. Hair provides warmth and protection for the skin. Eyebrow and eyelash hairs prevent particles of dust from entering the eye.

Teacher Preparation
Check schedule in advance; set aside time to complete five- to ten-minute "interviews" (**R**) with each child.

Materials
crayons	magnifying glasses	scissors	microscope (if available)
butcher paper	colored markers	**R: *About Me,*** page 9	

Activity: *What I Know About Me*
Use **R** on page 9 as an **"interview"** tool for each child.
During the interview, focus on the child's visible body characteristics, perceptions, and feelings. Although older children can fill in this page themselves, the interview format by a teacher or other classroom adult is preferable to reinforce the child's sense of individuality. Younger children can dictate responses to be written by the teacher.
Place each completed **R** in child's portfolio.

Name _____

About Me

My name is __John Doe__.
I am ___6___ years old.
My hair color is ___brown___.
The color of my eyes is ___green___.
The color of my skin is ___tan___.

Activity: *The Outside of Me*
Make a life-sized body outline picture of each child. Tape a large piece of butcher paper to a wall or to the

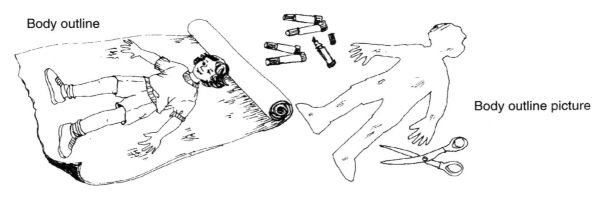

Body outline

Body outline picture

---

floor. Have each child stand or lie still against the paper, with arms extended away from body, and legs and fingers separated from each other. Use a heavy marker or crayon to trace around the child's body. Cut out the outline.

Students can use colored markers or crayons to draw hair, facial features, clothing, or other characteristics on their outlines.

Display body outlines. Save for later use. (The insides of the body can be colored in on the reverse side.)

Activity: My Hair

Where does hair grow? Examine hair on arms, legs, heads.

Under magnification:

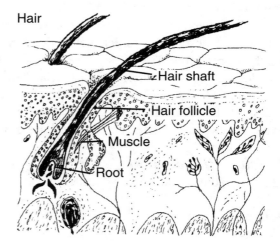

1) Cut or pull a hair from head to look at in detail under a magnifying glass or microscope. How does it compare with hair on arms in color, texture, thickness, and length?
2) Compare shape, size, and structure of root and shaft.
3) Compare/contrast curly, straight, and wavy hair.
4) Compare thicknesses of different head hairs.

Does hair color seem to influence the thickness or texture of the hair shaft? What else do students notice?

Recording

R: *About Me* completed and placed into portfolios.

Journals: Enter observations, discoveries, or impressions depicted in words, sentences, or drawings.

NOTES

About Me

My name is _____.

I am _____ years old.

My hair color is _____.

The color of my eyes is _____.

The color of my skin is _____.

My favorite thing to do outdoors is _____.

My favorite thing to do indoors is _____.

I am good at _____.

At night, I usually go to bed at _____.

In the morning, I usually wake up at _____.

My favorite food is _____.

One food I don't like is _____.

I feel happy when _____.

I feel sad when _____.

I get mad when _____.

My favorite school activity is _____.

My favorite storybook is _____.

I like to use my hands to _____.

I like to use my legs and feet to _____.

Turn this paper over to see a drawing I have made of me.

The Outside of the Body:
Skin & the Sense of Touch

Skin is the largest organ of the human body. It consists of a surface layer of dead cells that protects a layer of living cells below. Skin functions in several ways to protect the human body. It is a **sensory receptor** with nerve endings that can detect changes in the environment such as heat, cold, pressure, motion, itches, tickles, and pain. It **regulates body heat** through its numerous **sweat glands** that cool the skin through the evaporation of sweat. And it is a protective covering that **deters germs** from entering the body. Freckles and dark skin result from a pigment called **melanin** found in cells in the lower layers of the skin. Skin should be kept clean so that it functions as well as possible. Students have undoubtedly experienced cuts, scrapes, or other open sores that cause temporary breaks in the skin. Proper cleansing and bandaging help protect the sore areas from germs during the healing process.

Teacher Preparation
Place first-aid items in the display area for students to explore and use in dramatic play. Include bandages of various shapes and sizes, gauze, and so on. Discuss the purpose of each item with students.

Materials

magnifying glasses	butcher paper
ice cubes	feathers
stamp pads	paint rollers (brayers)
warm water	plastic cups
blank newsprint	poster paint
paper towels	**R:** *Fingers, Hands, Feet,* page 12

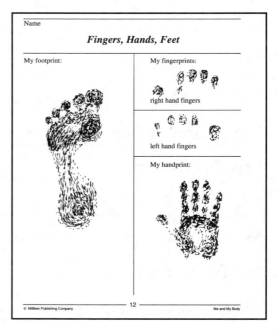

Activities: Wrinkles, Whorls, and Folds
Look at fingertips, palms of hands, soles of feet.

Compare lines in palms of hands. Are any two palms identical? Look at fingers, palms, and soles with a magnifying glass.

Use **R** on page 12. Make fingerprints of all ten fingers, using stamp pads. Are any two prints identical? Look at prints with magnifying glass. What patterns can be seen?

Use a paint roller (brayer) dipped in poster paint to spread a thin layer of paint over palms of hands and bottoms of feet. Make prints of hands and feet on **R.** (Use reverse side if needed.)

Make a classroom mural of students' hand-, foot-, and finger-prints.

Activities: Skin as Sensory Receptor and Temperature Regulator
Which part of the hand is most sensitive to touch? Use a feather to tickle fingertips, palm of hand, back of hand, wrist, web of skin between fingers. Which part is most sensitive to touch? Try this on the inside and outside of the elbow, and the front and back of the neck.

The Incredible Human Body

Hot or cold to the touch? Fill a plastic cup with very warm water. Fill a second cup with cold water containing ice cubes. Fill a third cup with water that is "skin temperature." Each student should hold a finger of one hand in ice water for several seconds and a finger from the other hand in warm water for several seconds. Then place both fingers into the "skin temperature" water. What happens? (To the "ice-water finger," the third cup of water will feel warm; to the "warm-water finger," it will feel cool.)

How does sweat cool our skin? Wet a paper towel. Rub it over one arm. Wave both arms in the air. Which arm feels cooler?

Going Further

Create a "detective story." Place painted finger-, hand-, or footprints in various parts of the room. Write individual stories, or a group story, based on clues discovered from the prints.

Create a "mystery box" into which small classroom objects can be placed. Students can describe and identify objects by reaching into the box and using only the sense of touch to guess what the objects are.

Create a fingerprint matching game by duplicating samples of students' fingerprints.

Interview the **school nurse** about caring for skin; discuss first aid for cuts, scrapes, and burns.

Scrape cheek cells from inside the mouth with a toothpick; view on a slide under a microscope.

Recording

R: *Fingers, Hands, Feet* completed and placed into portfolio.

Journals: Enter observations, impressions, and discoveries depicted in words, sentences, or drawings.

NOTES

The Incredible **Human** Body

Name

Fingers, Hands, Feet

My footprint:

My fingerprints:

right hand fingers

left hand fingers

My handprint:

The Incredible Human Body

The Outside of the Body:
Eyes & The Sense of Sight

The **human eye** is a sensory receptor of the **nervous system** that enables us to take in vast amounts of information about our world. "Seeing" happens when light rays from an object enter the eye through the **pupil** opening of the eye. The light rays pass through a **lens** which focuses a visual image onto the **retina** at the back of the eye. From there, specialized **nerve cells** transmit the image to the **brain** for interpretation. The size of the pupil opening is controlled by the **iris**, a circle of colored muscle surrounding the pupil. In dim light, the iris contracts and the pupil dilates, allowing more light rays to enter the eye. In bright light, the iris expands and the pupil becomes smaller. When students have their eyes examined, the eye doctor may artificially dilate the pupils with eyedrops. This allows more light rays to enter the eye and, under bright light conditions, everything looks somewhat blurred. Other parts of the eye include the transparent **cornea** covering the pupil and the iris; the white **sclera** surrounding the iris, the protective **eyelid** which closes over the eyes, **muscles** that control eye movements, a watery fluid in the space between the cornea and the lens, and a jelly-like substance between the lens and the retina. Good vision depends on many different factors. The shape of the eyeball, the shape of the cornea and the lens, and the action of muscles within the eye all play a role in determining how well the eyes move and are able to focus an image on the retina. Eyeglasses, contact lenses, and sometimes special eye exercises help compensate for some of the irregularities in the shape or function of parts of the eye.

Teacher Preparation

Try to locate a Snellen Eye Chart typically used by eye doctors; place it in the display area. Place several lenses in the display area for students to explore. Include convex and concave lenses, mirrors, and a magnifying mirror. Include instruments such as binoculars, camera, telescope, microscope.

Materials

small hand mirrors	newspapers	water (or mineral oil, light corn syrup)
small cups	a variety of small classroom objects	7.5 cm x 7.5 cm (3" x 3") pieces
eye droppers (optional)		of plastic wrap

Activities: Pupils and Lenses

Observe how the pupil of the eye changes size: Have students look into small hand mirrors with both eyes open, then close one or both eyes for several seconds. When students open eyes again, they can observe the action of their pupils in the mirrors. Repeat several times, with eyes closed for varying lengths of time.

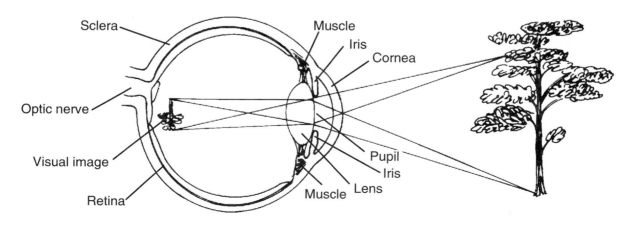

The Incredible Human Body

Observe what lenses can do: Give students ample opportunity to experiment with lenses in display area. As students move some lenses toward or away from their eyes, the image viewed may become inverted. This also happens with vision each time the lens of the eye focuses light rays from an object onto the retina. The image, which appears upside- down on the retina, is interpreted as right-side-up by the brain.

Make a water-drop lens: Students should work in pairs. One student holds a 7.5 cm x 7.5 cm (3" x 3") piece of plastic wrap taut, slightly above the print on a newspaper, while the second student places a drop of water (or a drop of mineral oil or light corn syrup) on the plastic wrap. (Drip liquid from an index finger if eye droppers are unavailable.) What happens to the size of the print viewed through the droplet? Carefully raise and lower the plastic wrap, keeping the droplet intact. What happens?

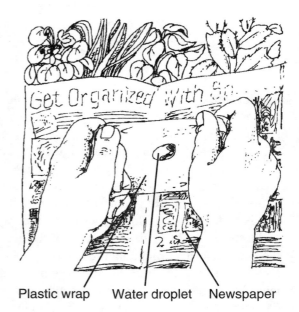

Plastic wrap   Water droplet   Newspaper

Activity: The Brain Remembers Visual Images

Place a variety of familiar classroom objects in a particular sequence on a table. Ask students to look at the objects for a few moments, then close their eyes. While eyes are closed, remove one of the objects. Students open eyes again and tell what is missing. Change the sequence, number, or grouping of objects; repeat.

Going Further

Find out about aids used by blind people: guide dogs, Braille, intensified use of the sense of touch.

Recording

Journals: Enter observations, discoveries, or impressions depicted in words, sentences, or drawings.

NOTES

The Incredible Human Body

The Outside of the Body:
Ears & the Sense of Hearing

Like the eye, the **human ear** is a special sensory receptor of the **nervous system**. The **outer ear** collects sound vibrations travelling in the air. The sounds move through the ear (auditory) canal and vibrate against the **eardrum** at the end of the canal. The eardrum then vibrates against three tiny, moveable bones within the **middle ear**. These bones, called the **hammer**, **anvil,** and **stirrup**, amplify the sound vibrations. The stirrup pushes against the **oval window** which separates the middle and inner ear, sending the sound vibrations to a coiled tube in the **inner ear** called the **cochlea**. Liquid inside the cochlea is set into motion by the sound vibrations. The liquid pushes against tiny hairs in the cochlea which are connected to sensory nerves in the **organ of Corti**, the sense organ for hearing located within the cochlea. The nerves pick up the cochlear "messages" and send them to the **brain** where they are interpreted as sounds.

The ear is also responsible for our sense of **balance**. Three fluid-filled **semicircular canals** within the inner ear carry out this function. One canal lies in a horizontal position; two are in vertical positions. When the body moves, fluid in the canals moves against tiny hair cells at one end of each canal. The hair cells stimulate nerves that carry messages about the body's position to the **brain** every time we sit, stand, and move up and down, side to side, and forward or backward. The brain sends messages back to muscles that must move to maintain the body's balance.

Teacher Preparation
Secure a tape recorder. Schedule time during the week to record each child's voice individually.

Materials
tape recorder
blank tape cassette
masking tape

shoeboxes (from home)
rubber bands, various sizes

12 identical empty juice cans with lids
small classroom objects (chalk, blocks, etc.)

Activity: The Sounds of Voices
Tape-record students' voices: Have each child say a few sentences into a tape recorder. After all voices have been recorded, play the tape to the entire group. Can students identify one another's voices? Their own voices? *Note: A child's tape recorded voice will sound different than normal speech.* During normal speech, sound vibrations reach the organ of Corti from *two* sources: the air, and the bones of the head. Taped sounds are carried through the air alone. Place fingers of one hand on the front of the neck, against the "voice box" (larynx); hum. Can students feel **vibrations?**

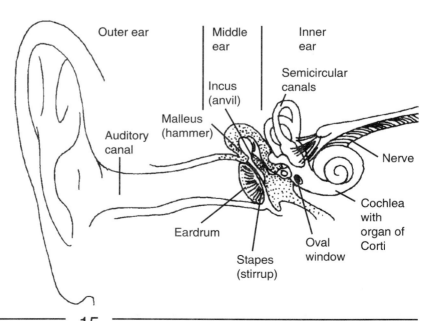

Outer ear

Middle ear

Inner ear

Semicircular canals

Incus (anvil)

Malleus (hammer)

Auditory canal

Nerve

Eardrum

Stapes (stirrup)

Oval window

Cochlea with organ of Corti

The Incredible Human Body

Activities: Creating "Soundmakers"

Make a shoebox guitar: Encircle an empty shoebox, cross-wise, with at least ten rubber bands of various sizes and thicknesses. Space the bands about 1.25 cm (½ inch) apart. Pluck or strum the "guitar." How does the thickness or tautness of a rubber band affect its rate of **vibration**? Which rubber bands have high sounds? Low sounds?

Make matching pairs of sounds: Enclose one each of pairs of small classroom objects—paper clips, chalk, unit blocks, puzzle pieces, and so on—in identical empty juice cans. Make at least six pairs: 12 juice cans with one object in each. Tape lids securely to each can. Place cans in display area where students can shake them, match sound pairs, and sequence sounds from loudest to softest.

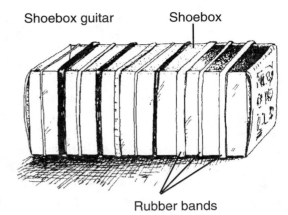

Shoebox guitar   Shoebox

Rubber bands

Going Further

Make tin-can "telephones" using two tin cans connected by string. Talk into one can; listen in the other.

Paint to sound. Explore the relationship between musical sounds and visual expression.

Find out about aids used by people with hearing loss: hearing devices, sign language, and lip reading.

Recording

Journals: Enter observations, discoveries, or impressions depicted in words, sentences, or drawings.

NOTES

The Incredible Human Body

The Outside of the Body:
Nose & Mouth, Smell & Taste

The **nostrils** of the **nose** capture smells carried in the air each time we breathe. The smells travel to tiny sensory (olfactory) nerves in the nasal membrane high up in the nose. The sensory nerves then send the smell "messages" to the **brain** for interpretation. Like the senses of touch, sight, hearing, and taste, the sense of smell helps us learn about the surrounding environment.

The **mouth** houses the **tongue, teeth,** and **gums.** It is the opening through which we take in food and water to nourish our bodies. An open mouth is also an alternative pathway for breathing. The mouth is surrounded by **lips** which work together with the tongue, teeth, air, and vocal cords to help us form the sounds of **speech.** **Taste buds** on the tongue contain sensory nerves that transmit taste sensations to the **brain** for interpretation. The taste buds work in conjunction with the sense of smell, allowing the brain to differentiate between different flavors and foods. The tongue distinguishes only sweet, sour, salty, and bitter tastes. Further distinctions between foods are made by the sense of smell.

Teacher Preparation
Find out if a dentist or dental hygienist would be willing to speak to the class.
Find out what the school lunch menu consists of on the day that you do smell and taste activities.

Materials

small hand mirrors	paper cups	herbs, spices, extracts	unsweetened powdered chocolate
toothpicks	sugar	lemon juice	small opaque containers with lids
cotton balls	salt	water	**R: *My Tongue,*** page 19

Activities: Tongues, Taste Buds, and Teeth

Use snack or lunch periods to test different foods and describe taste and smell sensations.

Look at the tongue with a hand mirror. Locate taste buds. Are they all the same size? Color? Do some areas of the tongue contain more taste buds than others?

Test taste areas of the tongue: Provide each student with four toothpicks and four paper cups that contain sugar water, salt water, lemon juice, and a small amount of unsweetened powdered chocolate. The students should use the toothpicks to place small amounts of each substance on different areas of the tongue: the tip, sides, middle, and back of the tongue. Can they tell which areas are most sensitive to which tastes? Use **R** on page 19 to record the taste areas of the tongue.

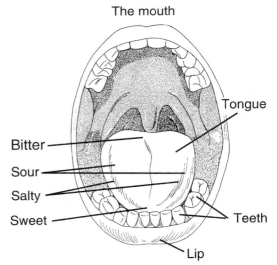

The mouth

Tongue

Bitter

Sour

Salty

Sweet

Teeth

Lip

Explore the role of teeth: In a large group, talk with students about the important role of their teeth. Share experiences about loose teeth, the loss of baby teeth, visits to the dentist, and care of the teeth.

Make a "tooth pillow" with a pocket in which to place a lost baby tooth for retrieval by the "tooth fairy."

The Incredible Human Body

Activity: Secret Scents

Use herbs, spices, extracts, or condiments to create secret scents. Moisten cotton balls with vanilla, cherry, lemon, peppermint, or chocolate extract. Use cinnamon, oregano, ketchup, minced garlic. Place each substance into a small, covered opaque container. Punch small holes in the covers to allow the scents to escape. Leave the scents in the display area for students to smell, describe, and compare. Ask them to bring a secret scent from home to share.

Going Further

Engage all five senses with one or more of the following food experiences: pop popcorn, make peanut butter by crushing peanuts, carve a pumpkin and toast its seeds, make applesauce.

Recording

R: *My Tongue* completed and placed into portfolios.

Draw in the **brain** on the back of the **body outline picture** from *The Outside of Me* activity (see page 7).

Journals: Enter observations, discoveries, or impressions depicted in words, sentences, or drawings.

NOTES

Name

My Tongue

Color the places on the tongue where you tasted sweet, sour, salty, or bitter tastes. Use a red crayon to color "sweet" areas, yellow to color "sour" areas, blue to color "salty" areas, and brown to color "bitter" areas.

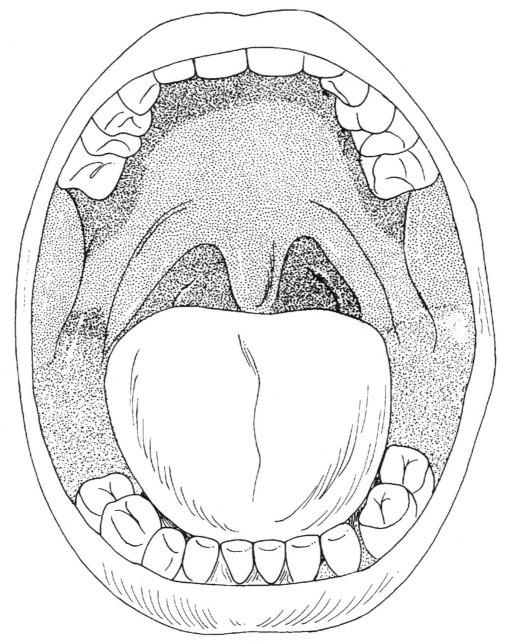

The Incredible Human Body

The Inside of the Body: *Bones & Muscles*

The **human skeleton** system consists of 206 bones that form a framework for the body, give it shape, and are the point of attachment for **muscles.** Bones protect vital organs such as the heart, lungs, brain, and spinal cord. Bones also contain **marrow** that produces blood cells. One bone joins another at a point called a **joint.** Bones bend at joints when **skeletal muscles** contract or expand. Skeletal muscles are attached to bones by tough tissues called **tendons.** Most skeletal muscles work in pairs in order to move bones. An example would be the action of triceps and biceps in the upper arm. When the forearm is raised, the biceps, a **flexor,** contract while the triceps, an **extensor,** relax. When the forearm is straightened, the triceps contract while the biceps relax.

An easy way to collect a supply of bones is to boil the carcass of chickens, turkeys, or fish after the meat has been eaten. After boiling, scrape the bones clean and let them dry overnight. Ask a local butcher to save large bones from other animals; boil and scrape these as well. If you are able to obtain animal skulls, students can examine the shape and size of teeth and observe how teeth fit into the jaw bone. As students study bones and muscles, they will become familiar with a variety of bones. Students can explore similarities and differences between bones in different parts of the body, discover that bones in the human body are similar in shape and location to those in other animals, and understand more about body movement. Students can bring bones from home to add to classroom collections to sort or assemble into skeletons.

Teacher Preparation
Assemble a large collection of bones from different animals. Place a variety of bones in display area.
Try to secure some X-ray pictures of broken bones from a clinic or hospital; display in a window.

Materials
bone collection	shallow cardboard boxes	chicken legs and wings	**T(A):** *Skeletal System*
glue	movable skeleton	tongue depressors	**T(B):** *Muscular System*
tape			

Activities: Bones
Give students ample time to explore bones in display area.

Sort bones: Provide shallow boxes into which bones of similar size, shape, or body location can be placed.

Examine human bones: Project the transparency, **T(A).** As students view it, ask them to feel various bones in their bodies as you point to the bones in **T(A).** Discuss shapes, sizes, and numbers of bones, especially in hands and feet. Move skeleton at "joints;" ask students to imitate the movements.

Inspect joints: Take apart whole chicken legs and wings. How do two bones move at the joint? Are all joints the same?

Arrange assorted bones into a skeleton: Place collections of bones into shallow boxes. Students can work in teams of four or five to assemble bones roughly into the shape of a skeleton. Glue arranged bones to the bottom of the box.

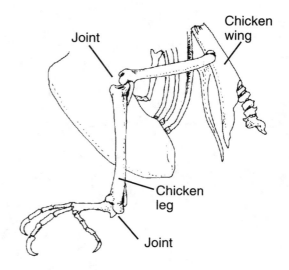

The Incredible Human Body

Explore broken bones: If X-rays of broken bones are available, tape them to a window for viewing. Talk with students about the need to immobilize a broken bone with splints, a cast, or a sling so that the broken parts can mend. Test a "make-believe" splint: use a tongue depressor and tape to immobilize a finger.

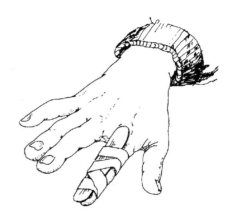

Activity: Muscles

Display **T(B).** Discuss the number and location of muscles pictured. Have students test action of their biceps: make a fist and raise the forearm; feel biceps with other hand.

Overlay the skeleton **T(A)** on **T(B).** Students can see how the bones and muscles move in conjunction with each other.

Test action of small and large muscles through movement activities.

Play different kinds of music while children interpret sounds and rhythms through body movements.

Recording

Draw as many bones (or muscles) as possible in their correct positions on the back of the **body outline picture** from *The Outside of Me* activity (see page 7).

Journals: Enter observations, discoveries, or impressions depicted in words, sentences, or drawings.

NOTES

The Incredible Human Body

The Inside of the Body:
Heart, Blood Vessels, & Lungs

The **circulatory system** consists of the **heart**, **veins**, **arteries**, and **blood**. The heart is a specialized muscle that contracts to pump blood throughout the body. The blood carries oxygen and food particles through arteries in order to nourish every part of the body. Blood travels through even smaller blood vessels, called **capillaries**, to reach every tissue. Blood also carries wastes such as carbon dioxide away from body organs, returning to the heart once again by way of the veins. The heart then pumps the blood to the **lungs** where it gives away carbon dioxide and picks up oxygen. The oxygenated blood returns to the heart, to be pumped out once again through the arteries. Each time the heart muscle contracts, arteries expand to make room for the oxygenated blood. This expansion or stretching can be felt as a **pulse** in arteries that lie close to the skin, such as in the wrist or temple. The pulse rate of a child is faster than that of an adult. The rate increases when a person is excited, scared, or has engaged in strenuous exercise.

The **respiratory system** allows the body to take in oxygen and expel carbon dioxide. The exchange of gases takes place through a network of capillaries in tiny **air sacs** within the **lungs.** When we breathe in, air travels from the nose or mouth into the **trachea** (windpipe), the **bronchial tubes**, and then into the air sacs of the lungs where oxygen passes directly into the bloodstream through the thin walls of the capillaries. Carbon dioxide passes from the capillaries into the air sacs and out of the body when we breathe out. The **diaphragm** muscle below the lungs contracts each time we breathe in, allowing the chest to expand, and relaxes again when we breathe out. **Hiccups** are spasms of the diaphragm muscle. Students may have heard that hiccups can sometimes be stopped by holding the breath for several seconds, thus keeping the diaphragm inactive.

Teacher Preparation
Obtain one or more stethoscopes to place in display area and to use in activities.

Materials

stethoscope(s)	clay	stopwatch	**T(C):** *Respiratory and*
toothpicks	balloons	paper, pencils	*Circulatory Systems*

Activities: Heartbeat and Pulse

Display **T(C).** Name the two systems pictured and the organs associated with each (see inside back cover); discuss.

Listen to the sound of the heart: Use stethoscopes; place on upper chest slightly to the left of center. The sound is that of valves within the heart snapping shut. Have students use a free hand to tap the rhythm of their heart sounds. Try listening to one another's hearts. Jump up and down several times; listen to the heart again. What has changed?

The Incredible Human Body

Find the pulse: Place the end of a toothpick in a small lump of clay. Place the clay on the inside of the wrist, just below the thumb. What happens to the toothpick? Time the pulse beats for one minute with a stopwatch; record the number of beats. Jump up and down several times; time pulse again for one minute; record results.

Activity: How We Breathe

Place hands firmly on skin just under the rib cage, then slowly take a deep breath and exhale again. Can students feel the action of the diaphragm muscle? Repeat. As students breathe in and out, the teacher can simultaneously blow into a balloon and let the air out to demonstrate how lungs fill with and expel air.

Have students exhale into one hand cupped around their mouths. How does the exhaled air feel? Warm? Moist?

Jump up and down several times. What happens to the rate of breathing?

Recording

Draw the heart, lungs, blood vessels, and diaphragm on the back of the **body outline picture** from *The Outside of Me* activity (see page 7).

Journals: Enter observations, discoveries, or impressions depicted in words, sentences, or drawings.

NOTES

The Incredible Human Body

The Inside of the Body: *The Digestive System*

Digestion is the process by which food is broken down into particles small enough to be absorbed by the bloodstream. **Digestive juices** containing **enzymes** help break the food down the moment it enters the **mouth**, where it is chewed and mixed with **saliva**. When the food is swallowed, it passes through the **esophagus** and into the **stomach**. It remains in the stomach for several hours where it is mixed with digestive juices that break it down further. During this time, the food is mixed by a churning motion caused by the contraction of strong stomach muscles. The food then passes into the **small intestine**. Here several more digestive juices, produced by the **liver**, the **pancreas**, and the walls of the small intestine, complete the process of digestion so the food can be absorbed into the bloodstream. Capillaries within the wall of the small intestine absorb the digested food particles. The **large intestine** or **colon** stores waste materials and undigested parts of food, eventually eliminating them from the body through the **anal opening.**

The foods we eat contain sugars, fats, proteins, and starches, each of which is acted upon by **enzymes** or acids in the digestive juices. Sugars are usually digested quickly, while fats take the longest to be digested. You might want to include information and activities about nutrition (see page 28) during the discussion of digestion.

Teacher Preparation
Ask the school dietician or head of cafeteria operations to give you information about the sugar, fat, protein, or starch content of a particular day's lunch menu. Share this information with students.

Materials
crackers	a plain white t-shirt	classroom clock
cold juice or water	heavy red marker	**T(D):** *Digestive Tract*

Activity: What Happens to Our Food?
Display **T(D).** Name the organs pictured (see inside back cover). Discuss the digestive system. What do students know?

Place crackers and cold juice or water on the table for each student. Ask students to swallow some juice. Can they feel the cold liquid moving down the esophagus toward the stomach? Take another swallow of juice while watching the second hand on the clock. How long does it take for the juice to reach the stomach? Count seconds until students no longer feel internal movement of juice.

Next, test the crackers. Which teeth are used to bite the crackers? To chew them? Do students feel the crackers getting mushy as they chew? When the crackers are swallowed, can students feel them moving down the esophagus toward the stomach? How long do the students think the crackers will remain in the stomach? Talk about what will happen to the crackers in the stomach, where the particles of the crackers will go next, and approximately when the crackers will be digested.

Make a chart that shows what happened when the students ate the juice and crackers.

EATING JUICE & CRACKERS

11:00 a.m. – Swallowed juice.
4 seconds later – juice goes into stomach.
11:02 a.m. – Ate crackers.
Crackers moved down esophagus into stomach.
Stomach mixed crackers and juice with digestive juices.
Food moved into small intestine.

The Incredible Human Body

Activity: Make a Classroom Prop for Dramatic Play

Use an old white t-shirt and a heavy red marker. Draw the esophagus, stomach, and large and small intestines on the front of the shirt. Place in the display area for students to use. Students may want to bring old t-shirts from home to create their own "digestive" t-shirt.

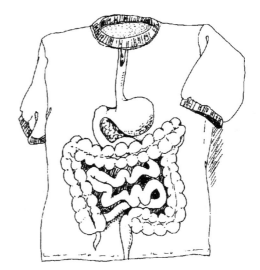

Going Further

Test small bits of food placed on the tongue to see which foods saliva dissolves faster than others.

Recording

Draw the organs of the digestive system on the back of the **body outline picture** from *The Outside of Me* activity (see page 7).

Journals: Enter observations, discoveries, or impressions depicted in words, sentences or drawings.

NOTES

The Expressive Body: *Sensations & Emotions*

The human body monitors its own internal environment through the sensations of **hunger, thirst, pain,** or **fatigue.** Sometimes called the **internal senses**—as opposed to the external senses of sight, hearing, touch, smell, and taste—these sensations are produced by chemical or physical changes in organs and systems of the body and are relayed by nerve cells to the **brain.** The brain sends back appropriate responses to keep the internal environment stable.

Emotions are reactions to thoughts or external circumstances that cause a person to feel angry, sad, happy, or fearful. Emotions can protect the body by triggering the release of **hormones** that allow the body to respond to danger. For example, if a child is confronted by a snarling dog, the emotion of fear may cause adrenalin to be released from the adrenal glands into the child's bloodstream. The adrenalin then triggers increases in heartbeat, blood pressure, and rate of breathing. Blood shifts from the digestive system to the muscles and the brain and more sugar enters the bloodstream. These changes, sometimes called the "fight or flight response," give the body energy to respond to the emergency. Certain emotions cause body reactions such as "goose bumps," dilated pupils, tears, and laughter. Emotions are often reflected in a person's facial expressions, posture, body gestures, or tone of voice. People decide how to respond to one another by observing these outward signs of the emotions.

Teacher Preparation
During a two to three day period, take notes about student reactions to various classroom stimuli, including new and old routines, outdoor activities, meal time, and so on. Note verbal and non-verbal responses about hunger, pain, thirst, or fatigue. Note emotional expressions and the circumstances that provoked them.

Materials
teacher notes (see paragraph above)
easel with pad of large blank newsprint
colored markers

Activity: The Internal Senses
Discuss with students some of the observations from your notes. These might include requests to go to the drinking fountain, or statements such as, "I'm still hungry," or, "I'm too tired."

Name the internal senses; discuss their importance.

Create an "Internal Senses" chart: Use blank newsprint and a colored marker. List the internal senses across the top of the chart. Ask students to describe what happens when they feel tired (yawning?), thirsty (dry mouth?), hungry (grumbling stomach?), or when something hurts (soreness?). Write their descriptions on the chart. What actions do they take to alleviate feelings created by their internal senses? Again, write responses on the chart.

	Hungry	Thirsty	Tired	Feel Pain
How I Know	Hunger Pains Grumbling Stomach	Dry mouth Feel hot	Yawn Eyes droop Cranky	Sore Something hurts
What I Do	Eat something	Get a cold drink of water	Lay down Take a nap Rest	Tell the teacher Tell my mother

The Incredible Human Body

Activities: The Emotions

Discuss with students some of the observations from your notes regarding emotional responses and the circumstances that triggered them.

Discuss why the emotions are important.

Ask students to describe some of the emotions they have experienced themselves or observed in others. Make a **list of terms** students use to label emotions. *Note: In developing a list of terms, include words that describe **degrees** of emotional responses, such as sad, grief-stricken; irritated, angry; happy, overjoyed.*

Pantomime emotional expressions: Each student can use face/body gestures to depict an emotional response. Can others identify the emotion?

Draw facial expressions: Draw a series of circles ("faces") on blank newsprint. Describe a situation that would provoke an emotional response. Students can draw in the facial features that express it. Give each drawing a label according to the emotion it expresses.

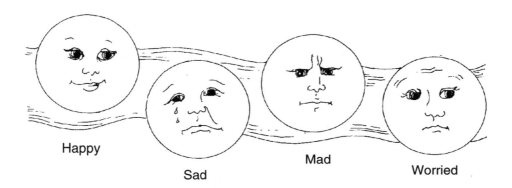

Happy Sad Mad Worried

Recording

Journals: Describe "How I felt when . . ." and "What I did . . ." as a result. Or, draw pictures that show the first feeling, then the second one.

NOTES

The Incredible Human Body

Caring for Our Bodies:
Nutrition, Health, & Safety

Nutrition plays a vital role in the growth, development, and maintenance of the human body. Food is the "fuel" which the body converts into the energy it needs to build cells and tissues, to repair or replace injured or worn-out tissues, and to maintain all of the body's functions. Children require more food energy than adults because their bodies are growing. The energy in any given food is measured in **calories**. Food itself consists of **nutrients**, which can be divided into five groups: **fats** and **carbohydrates** (sugars, starches), which provide the main source of energy in food; **proteins**, needed for growth and maintenance of bones, muscles, and other organs of the body; **vitamins**, which do not provide fuel but are essential for the health and growth of the human body; and **minerals**, which are fundamental parts of teeth and bones, body fluids, and digestive juices. A balanced diet should include one or more daily servings from the **five food groups of the food pyramid** to provide the necessary nutrients: **1) milk and dairy group, 2) grain group, 3) vegetable group, 4) fruit group, and 5) legumes and meat group** (see *Nutrition Chart,* page 30). **Water**, the basis of blood and other body fluids, is not really a food but is essential to human life and must be part of the daily diet.

In addition to a balanced diet, the human body requires daily exercise, rest, and cleanliness in order to maintain good health. **Exercise** helps build muscles and benefits circulation, which brings nourishment to all parts of the body. **Rest and sleep** allow the body to slow down in order to repair or replace worn-out cells. Injury from accidents requires the body to work even harder to repair damaged body parts. Some accidents can be prevented by observing **safety rules. Cleanliness** protects the body from germs that may cause disease. (See also *Bones & Muscles,* pages 20–21, and *Skin & the Sense of Touch,* pages 10–11.)

Teacher Preparation
Place empty boxes and cans of food, labels intact, in display area. Find out what foods are included in the school lunch menu during a two to three week period.

Materials
large pieces of blank newsprint,
   or poster board
empty food containers, with labels
R: *Nutrition Chart,* page 30

colored markers
paper, pencils

Activity: Nutrition
Talk with students about the body's need for food.
   What do they eat when they are hungry? Show nutritional labels on empty boxes or cans of food. Discuss the meaning of **nutrient** and of **calorie**. Encourage students to bring empty containers of food from home to add to the class collection. Use nutritional information on labels to decide which foods provide more "nutrient value" than others. Write those foods on a list to keep in display area. Which foods have the *least* nutritional value?

The Incredible Human Body

Recording nutritional intake. Make at least five copies of **R:** *Nutrition Chart,* on page 30 for each student. Students should label the day on the Nutrition Chart. Students should write down or draw pictures of the lunch foods they have eaten that day in the spaces provided on the Nutrition Chart.

Total the number of each kind of food consumed. Record the name of the food and the total number of servings eaten in the appropriate food-group column on the chart; or record it by pasting a picture of the food in the appropriate column.

Be sure to discuss the meaning of **"junk food"** with the students. Junk food usually contains a much greater proportion of fats and/or sugars, and little if any vitamins, proteins, or minerals. Although such foods have some nutritional value, they should not be habitually substituted for daily servings of other foods essential to meeting the body's nutritional needs.

At the end of a week, discuss the number and kinds of foods the class as a whole consumed during lunch.

Activity: Health and Safety Rules

Ask a school nurse to speak to class about good health habits.

Review the daily class schedule. The school day provides time to exercise, rest, wash hands, and eat. Students use energy from the food they eat to complete tasks required during a typical school day. Ask students about the various ways in which they use and take care of their bodies during the school day. Record responses on poster board or blank newsprint.

Create a health and safety poster that lists ideas and rules generated by students.

Going Further

Find out about community agencies or organizations that help protect and maintain good health.

Recording

R: *Nutrition Chart* completed and placed into portfolios.

Journals: Make daily entries about observations, discoveries, foods eaten, health habits, or decisions.

NOTES

The Incredible Human Body

Name

Nutrition Chart

In the chart below, draw pictures or write the names of the kinds of foods you eat at lunch time. Write the day here. _____

Milk and Milk Products 	Meat, beans and nuts 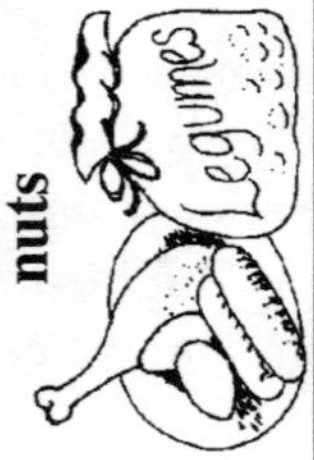	Breads, Cereals 	Fruits 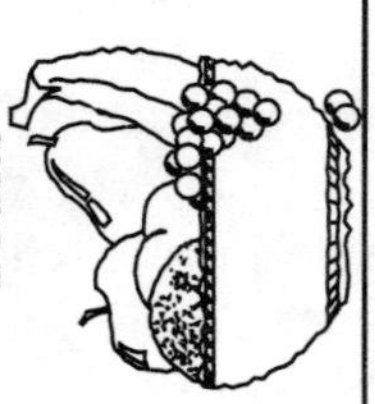	Vegetables 	"Junk Food" 

The Incredible Human Body

Application of Concepts: *Making Connections*

Portfolios and journals completed by each child provide a profile of observations, discoveries, and impressions about the human body from several individual points of view. Review each child's perceptions and observations before pursuing one or more of the activities suggested below. The activities provide the opportunity for each child and the whole class to consolidate their learning of both concepts and facts. Choose a project(s) which gives a sense of understanding, a sense of closure, and a real sense of enjoyment for your group.

1) Interview a school or community health worker. Ask for student input in creating a list of interview questions. Ideas might be: "What do you do when someone breaks an arm after falling at school?" "Does it hurt to have an X-ray?" "Who will help me out if I get sick at school?" "What happens when someone goes to the hospital?" "Why do I have to get shots before starting school?" Bring along a tape recorder to tape the interview. Place the tape and recorder in the display area for students to listen to again on their own.

2) Construct a life-sized "human body." Use copper wire, construction paper, glue, tape, poster paint, pipe cleaners, papier-mâché, clay, paste, or other materials. Make "skeletal bones" by cutting life-sized bone shapes from paper and gluing or taping copper wire around the perimeter of each paper bone. The wire will provide a stiff framework around which to build the rest of the body. Cut out "organs" from construction paper, paint them, and anchor them inside the "skeleton" with more copper wire. Use red and blue pipe cleaners to create "veins and arteries." Use students' ideas for materials that can help create the "body." Try to include as many parts of the body as possible.

3) Find out more about the brain and nervous system. The brain has been an integral part in the functioning of every organ and system students have studied. It has been an essential part of every activity in which the students have participated. In addition to responding to messages it receives from every part of the body, the brain is also the seat of learning, memory, intelligence, judgment, creativity, imagination, and thought. Find out more about the things the brain can do. Play memory or logic games in class. Tell a **"guided imagery"** story while students have their eyes closed, imagining the various settings or activities you describe in the story. Encourage creative thinking to solve problems in school activities. Ask small groups of students to research one facet of the brain's function using books, interviews, observations, or discussions with classmates. Results of research can be shared with classmates through oral or written reports.

4) Stage a play about the human body. Students can select "characters" based on organs or systems of the body. They can work in groups or teams to create simple costumes, develop a script based on functions of body parts, construct stage props, and select appropriate music. Other classes may be invited to view the play.

5) Invent a better toothbrush, a more efficient way to get nutrients to all parts of the body, a better bed for sleeping, a more effective set of teeth, a brain with even more functions, and so on. The idea is to let the students use their imagination and creativity to draw, design, or construct something that will allow them to take even better care of their bodies, or to enhance the function of one part of their bodies. Be prepared for a fun and inventive adventure!

6) Complete a class book. Ask students to select which pieces of information to include. They might want to submit copies of portfolio or journal entries, a written paragraph about themselves, or a summary of some of their activities. Include charts, photos of the children, drawings, and other information the class thinks is important, such as health and safety rules during the school day. Develop a list of "fascinating facts" the students have learned about their bodies. Create a book title, cover, table of contents, and list of "authors." Set aside a special place where the book can be displayed and used.

Bibliography

FOR CHILDREN
General
The Magic School Bus Inside the Human Body—J. Cole
    (Scholastic Books)
Slim Goodbody: The Inside Story— J. Burstein (McGraw-Hill)
You: How Your Body Works—L. McGuire (Platt & Munk)
Bodies—B. Brenner (Dutton)
Inside You and Me—C. L. Fenton & E. F. Turner (John Day
Co.)
All About Me—D. Manley (Raintree Publishers)
From Head to Toes: How My Body Works—M. Packard
(Messner)
People—P. Spier (Doubleday)

The Outside of the Body
My Five Senses—Aliki (Crowell)
Your Five Senses, A New True Book—R. Broekel
    (Children's Book Press)
Faces—B. Brenner (Dutton)
The Look Book —J.B. Moncure (Children's Book Press)
Find Out by Touching—P. Showers (Crowell)
Your Skin and Mine—P. Showers (Crowell)
How You Talk —P. Showers (Crowell)
Look at Feet—H. Pluckrose (Franklin Watts)
My Feet—Aliki (Harper Collins)
My Hands—Aliki (Harper Collins)
Look at Hands—R. Thomson (Franklin Watts)
Follow Your Nose—P. Showers (Crowell)
What Your Nose Knows!—J. B. Moncure
    (Children's Book Press)
Sounds All Around—J. B. Moncure (Children's Book Press)
High Sounds, Low Sounds—F. Branley (Crowell)
The Listening Walk—P. Showers (Crowell)
Ears Are for Hearing—P. Showers (Crowell)
How Many Teeth?—P. Showers (Crowell)
How Peter Molar Looked for a Smile—J. & S. B. Bohanak
    (Adonis Studio)
Straight Hair, Curly Hair—A. Goldin (Crowell)
Look at Hair—R. Thomson (Franklin Watts)

The Inside of the Body
A Drop of Blood—P. Showers (Crowell)
Hear Your Heart—P. Showers (Crowell)
What Happens to a Hamburger?—P. Showers (Crowell)
The Skeleton Inside You—P. Balestrino (Crowell)
Use Your Brain—P. Showers (Crowell)
Your Brain & Nervous System, A New True Book
    L. J. LeMaster (Children's Book Press)
Brain Power!—P. Sharp (Lothrop, Lee & Shepard)
The Human Body: The Brain—K. Elgin (Franklin Watts)
The Brain: What It Is, What It Does—R. D. & B. Bruun
    (Greenwillow Books)
About Your Brain—S. Simon (McGraw Hill)
Oxygen Keeps You Alive—F. Branley (Crowell)
Why I Cough, Sneeze, Shiver, Hiccup & Yawn—M. Berger
    (Crowell Junior Books)
Movement—J. Gaskin (Franklin Watts)

The Expressive Body
Feelings—Aliki (Greenwillow Books)

Caring For Our Bodies
Food Is For Eating—I. Podendorf (Children's Book Press)
Your Body Fuel—D. Balwin & C. Lister (Bookwright Press)
Clean and Strong—C. Brownell & J. F. Williams
    (J. B. Lippincott)

True Book of Health—O.V. Haynes (Children's Book Press)

FOR TEACHERS
General
Facts About the Human Body—M. & M. A. Tully
    (Franklin Watts)
*You Can't Sneeze with your Eyes Open, & Other Freaky
    Facts about the Human Body*—B. Seuling
    (Lodestar Books—E. P. Dutton)
Cuts, Breaks, Bruises & Burns: How Your Body Heals—
    J. Cole (Crowell)
Skin and Bones—J. E. DeJonge (Baker Book House)
Science in a Nutshell—I. Follman & H. Jackson (Kimbo)
You & Your Body—A. E. Klein (Doubleday)

The Outside of the Body
The Story of Your Eye—W. Hammond
    (Coward, McCann & Geoghegan)
The Story of Your Hand—A. & V. B. Silverstein
    (G. Putnam's Sons)
Hands Up!—A. Kramer (Macmillan)
The Story of Your Foot—A. & V. B. Silverstein
    (G.P. Putnam's Sons)

The Inside of the Body
The Human Brain—L. Kettelkamp (Enslow Publishers)
The Story of Your Brain and Nerves—E. L. Weart
    (Coward-McCann, Inc.)
Your Muscles—And Ways to Exercise Them —M. Cosgrove
    (Dodd, Mead & Co.)
Bones—H. S. Zim (William Morrow & Co.)
The Human Body: The Digestive System—K. Elgin
    (Franklin Watts)
Human Anatomy for Children—I. Goldsmith (Dover)

Caring For Our Bodies
Food and Nutrition—W. Sebrell & J. J. Haggerty
    (Life Science Library—Time, Inc.)
The Chemicals We Eat and Drink—A. & V. B. Silverstein
    (Follett Publishing Co.)
Your Food and You—H. S. Zim (William Morrow & Co.)
Your Health and You—C. Gramet (Lothrop, Lee & Shepard)

OTHER RESOURCES
Community resources to consider using:
    libraries
    museum of science (education division)
    local health clinics (information, materials)
    school health workers, cafeteria dietitians
    grocery stores
        (nutritionists, for information about foods)
        (butcher, for animal bones)
    zoo (education division, for animal bones or skulls)
National Dairy Council
    To request a catalog of materials for pre-K and
    grades 1–6, call 1-800-426-8271
    Or write:
        National Dairy Council
        6300 North River Road
        Rosemont, IL 60018
Children's Book Council
    To request a current list of Outstanding Science Trade
    Books for Children, write:
        Children's Book Council
        568 Broadway
        New York, NY 10012

The Incredible Human Body